COLORS & MARKINGS OF THE

F-14 TOMCAT

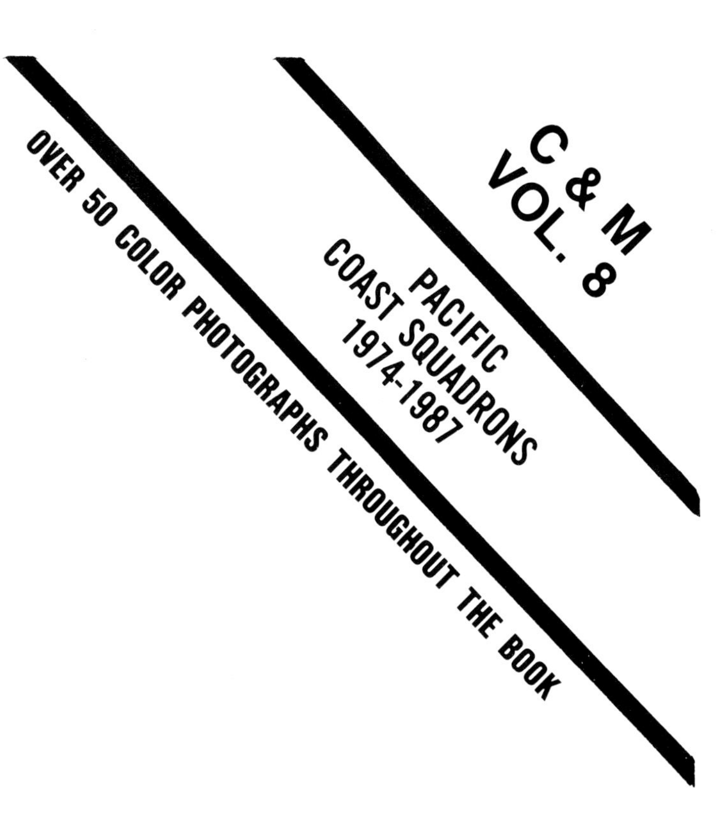

Bert Kinzey and Ray Leader

Distributed by:
Airlife Publishing Ltd.
7 St. John's Hill, Shrewsbury SY1 1JE, England

Airlife Publishing Ltd.
England

This book is a product of Detail & Scale, Inc., which has sole responsibility for its content and layout, except that all contributors are responsible for the security clearance and copyright release of all materials submitted. Published and distributed in the United States by TAB BOOKS Inc., and in Great Britain and Europe by Airlife Publishing, LTD.

CONTRIBUTORS:

Flightleader
Jerry Geer
Mick Roth
Mike Grove
Dwayne Kasulka
Peter Bergagnini
Brian Rogers
George Cockle
Don Logan
Matt Geer
Phillip Huston
Donald S. McGarry
Arnold Swanberg
Ben Knowles
Bob Stewart
Don Spering/AIR
Don Linn
Picciani Aircraft Slides
Tom Brewer
David Brown
John Binford
Steve Miller
Frank Nuanez Jr.
Doug Slowiak
GB Aircraft Slides
Craig Kaston
LT David Baranek, (VF-124)
LTJG Anne Cline, (VX-4)
LT Russell Sklenka, (VF-213)
LT Steve Pollard, (VF-213)
LT Robert Alcala (VF-213)
LTJG Ray Jalette (VF-194)
LTJG Kolin Campbell (VF-194)

FIRST EDITION
FIRST PRINTING

Published in United States by

TAB BOOKS Inc.
P.O. Box 40
Blue Ridge Summit, PA 17214

Library of Congress Cataloging
in Publication Data

Kinzey, Bert.
Colors & markings of the F-14 Tomcat.

(pt. 1: C & M ; vol. 2) (pt. 2: Color & Markings ; vol. 8)
Pt. 2: F-14 Tomcat / by Bert Kinzey and Ray Leader.
Published: Blue Ridge Summit, PA : Tab Books.
Contents: pt. 1. Atlantic coast markings the first ten years, 1974-1984 — pt. 2. F-14 Tomcat.
1. Tomcat (Jet fighter plane) I. Leader, Ray.
II. Title: Colors and markings of the F-14 Tomcat.
III. Title: F-14 Tomcat. IV. Title: F-fourteen Tomcat.
V. Series: C & M ; vol. 2. VI. Series: Colors & markings
ISBN 0-8168-4526-3 (pbk. : pt. 1)
ISBN 0-8036-8032-2 (pbk. : pt. 2)

First published in Great Britain in 1987
by Airlife Publishing Ltd.
7 St. John's Hill, Shrewsbury, SY1 1JE

British Library Cataloging in
Publication Data

Kinzey, Bert
F-14 Tomcat. (Colors & markings series)
1. Tomcat (Jet fighter plane)
2. Airplanes, Military — Identification marks
I. Title II. Leader, Ray III. Series
623.74'64 UG1242.F5
ISBN 9-85368-838-9

Questions regarding the content of this book should be addressed to:

Reader Inquiry Branch
Editorial Department
TAB BOOKS Inc.
P.O. Box 40
Blue Ridge Summit, PA 17214

Front cover: Both colorful and low-visibility markings for VF-111 are seen in this painting by aviation artist Dallas Lloyd. The painting was done specifically for the cover of this book. The "Sundowners" markings are among the most popular ever carried on the Tomcat.

Rear cover: This nice in-flight photograph shows one of VF-213's Tomcats that was painted in a temporary camouflage pattern. The photo was taken over the California desert. (LT Pollard)

INTRODUCTION

Probably the most unusual and short-lived paint scheme ever applied to any Tomcat is this scheme used during a VX-4 change-of-command ceremony. It is similar to a scheme used previously on an F-4J, and later an F-4S, that belonged to VX-4. The aircraft is overall black with the usual VX-4 blue bands on the vertical tails. These bands are outlined in gold, and have white stars. A white ***XF*** *tail code is within the bands on the outside of the tails. The most dominate feature is the large, white Playboy bunny on the tail, just as it appeared earlier on the Phantoms.* ***CAPT MOON VANCE*** *appears on the forward left canopy rail. The aircraft is F-14A, 158358, but no number was painted in the usual location on the engines or on the tail. It seems that the scheme and markings were very popular with everyone except an admiral who expressed displeasure about it. The water-soluble paint was removed after only two days. (VX-4)*

F-14 Tomcats that are based on the Atlantic coast were the subject of Volume 2 of the Colors & Markings Series. In this book, Detail & Scale concludes the coverage of the Tomcat with a look at all of the squadrons based on the Pacific coast. All fleet and reserve squadrons that are based at NAS Miramar, California, are presented in numerical order, and are followed by the aircraft of VX-4, the Naval Missile Center, Pacific Missile Test Center, and NASA. Even the Tomcats that were specially marked and used in the movie "TOP GUN" are included. For each squadron, coverage begins with the earliest markings and scheme used, and concludes with the present colors carried by that squadron's aircraft.

The presentation of photographs in this publication is improved over that in the Atlantic coast book, because in this book the color photographs are located throughout all of the pages instead of being located in one color section in the center of the book. This provides much better continuity in the presentation, and makes for a nicer looking publication.

In most of the volumes in the Colors & Markings series, coverage is presented on a unit-by-unit basis. This requires extensive research and a large selection of photographs to work from in order to provide a good look at the often widely varied paint schemes and markings used by each unit. While it is relatively simple to gather a collection of photographs and organize them into a book, it is quite another matter to cover all the bases when every unit and all of their major markings are to be included. Such a book as this cannot be done without the help of a lot of contributors, and their assistance is acknowledged here. Their names appear on the previous page, and the list is rather extensive. A special thanks is due PAOs LTJG Anne Cline of VX-4, and LT Russell Sklenka of VF-213. Their efforts provided several unusual photographs that could not have been found elsewhere. The authors would also like to express their thanks to Dave Hornick and Dave Legg of the Aircraft Combat Survivability Branch, for providing information on paint schemes, and to Roy Grossnick of the Office of Naval Aviation History and Archives, for providing background history and information on each of the squadrons. Lastly, the efforts of Phillip Huston should be recognized. Phillip spent a lot of time and effort to photograph the first aircraft and first markings to be used by VF-191 and VF-194 on their new Tomcats. The photographs of the first F-14 to be marked by VF-191 came at the very last minute, and without Phillip's help, as well as the cooperation of the squadron itself, we would not be able to include photographs of that unit's first F-14 markings. Jerry and Matt Geer also provided last minute photographs of these two squadrons.

The choice of photographs was made carefully so as to illustrate as many different schemes for each unit as possible. While it is impossible to show each minor variation of schemes used by every unit, the photographs on the following pages do present the most complete look at the markings used by these squadrons ever presented in a single publication. As an interest to the historian, dates that the photographs were taken are given whenever possible. Generally, the photographs for each squadron are arranged in a chronological order.

F-14 PAINT SCHEMES AND MARKINGS

This F-14A, 159608, fro. , VF-211, exemplifies the gray over white scheme and colorful markings that were in vogue when the Tomcat entered service and for a number of years thereafter. Red and white checks appear on the rudders, the ventral fin is red, white, and blue, and stripes of the same colors are on both sides of each tail. Red and white stripes outline the black anti-glare panel. Control surfaces are white, and large, colorful national insignias are used. This is a far cry from the tactical scheme and markings used today. *(Flightleader)*

Although Colors & Markings Volume 2, which covered the Atlantic coast F-14 squadrons, provided a summary of the paint schemes used on the Tomcat, it is worth repeating here. Additionally, in the three-year time period that has elapsed since that book was published, the tactical paint scheme has become the standard for F-14s, and more information is provided here on that scheme.

When the Tomcat entered service, the gray over white scheme was the standard for Navy fighters and several other types of aircraft. This scheme consisted of light gull gray (FS 36440) on the upper surfaces, and gloss white on the under surfaces. The control surfaces, such as rudders, spoilers, flaps, and stabilators, of these aircraft were white, as were the interiors of the wheel wells and intakes. Landing gear struts and wheels were also white. This scheme was usually adorned with extensive amounts of color that was applied by the using squadrons.

When the move to low visibility began, the white was deleted from the gray over white scheme, and the aircraft were painted overall light gull gray. Only the wheel wells and struts remained white. At first there was still a considerable amount of color used on this scheme, but gradually this became less and less as even the national insignias became gray or black. Most unit markings also were gray or black, as were the standard stencils and rescue markings. Various amounts of color could be seen on the overall gray scheme, but generally it became less and less as time went on. Today a few Tomcats remain in the overall gray scheme, but it has largely been replaced by the tactical blue/gray scheme.

The official tactical scheme for the F-14 consists of three shades of blue/gray. The entire top of the aircraft is FS 35237, which is the darker of the three shades. The bottom of the aircraft is painted in the lightest shade, which is FS 36375. The sides of the nose section of the fuselage back to the intakes and the vertical tails are painted FS 36320. In the official scheme, there is no change in color for the area behind the cockpit or the leading edges of the wings, as is the case for the A-6 and A-7. The idea is to countershade with the three grays so that, regardless of the lighting conditions and attitude of the aircraft, it will be difficult to see. Officially, the markings are to be contrasting shades of the three grays. For example, on the darker 35237 uppersurfaces, 36375 would be used for the markings. The one exception is the modex. Here the unit may use flat black or engine gray (FS 36081). There is no color whatsoever officially on the tactical scheme, although some units have used it spar-

The first step to subdued, low visibility schemes and markings was to delete the white in the gray over white scheme, and paint the aircraft in overall light gull gray. While some color often remained on this scheme, much of it was eliminated. This F-14 from VF-124 shows the use of the overall gray scheme and low visibility markings. The only color that remains is the large red, white, and blue national insignia. All other markings are dark gray on this aircraft. *(Cockle)*

The official scheme used on Tomcats today is the tactical blue/gray scheme as seen on this aircraft from VF-302. Absolutely no color remains except in rare instances. Some markings are almost impossible to see unless the observer is very close to the aircraft. *(Flightleader)*

ingly.

When an aircraft comes out of the rework facility, it should be painted in this official scheme, but once the aircraft is at the unit, there may be considerable variation in the painting and repainting that is done to it. F-14s can be seen in just one of the three grays, or in only two of them. Once the aircraft is spot painted for corrosion control purposes, it may be difficult to tell how many shades are used and exactly where they are. This explains the deviations from the official scheme that are seen on the following pages of this book.

In the past few years there has been some experimentation with some rather unusual camouflage patterns on the Tomcat. One of these appears on the rear cover, and others can be seen in several places within the book. These camouflage schemes are usually temporary, and are applied with water soluble paints that are supposed to be easy to remove. They are rather crude, often being applied with large brushes, mops, or even brooms. A few were sprayed on with an air gun. These schemes are usually worked out on the spot, being completely unofficial, and result in some unique appearances for any fighter in the entire history of U.S. Naval Aviation.

The Pacific Fleet F-14 fighter squadrons are based at NAS Miramar, California. Two squadrons are assigned to each carrier air wing, and each air wing is assigned to a carrier. These assignments are changed from time to time as carriers go into the yard for extended maintenance periods, and for other reasons. The assignment of each fleet squadron is given below, and is current as of the press date for this book, which is July 1987. Additionally, VF-124 is the Fleet Readiness Squadron, and is not assigned to an air wing or carrier. VF-301 and VF-302 are reserve squadrons assigned to Reserve Air Wing 30 (CVWR-30). They do not have a carrier assignment. It should also be noted that one Pacific carrier, **USS Midway, CV-41,** does not operate F-14s, and therefore does not show up in the table below.

UNIT	NICKNAME	CARRIER AIR GROUP	CARRIER
VF-1	Wolfpack	CVW-2	USS Ranger (CV-60)
VF-2	Bounty Hunters	CVW-2	USS Ranger (CV-60)
VF-21	Freelancers	CVW-14	USS Constellation (CV-64)
VF-24	Fighting Renegades	CVW-9	USS Kitty Hawk (CV-63)
VF-51	Screaming Eagles	CVW-15	USS Carl Vinson (CVN-70)
VF-111	Sundowners	CVW-15	USS Carl Vinson (CVN-70)
VF-114	Aardvarks	CVW-11	USS Enterprise (CVN-65)
VF-124	Gunfighters	RAG	None
VF-154	Black Knights	CVW-14	USS Constellation (CV-64)
VF-191	Satan's Kittens	CVW-10	USS Independence (CV-62)*
VF-194	Hellfires	CVW-10	USS Independence (CV-62)*
VF-211	Fighting Checkmates	CVW-9	USS Kitty Hawk (CV-63)
VF-213	Black Lions	CVW-11	USS Enterprise (CVN-65)

* The USS Independence is presently undergoing the Service Life Extension Program (SLEP). When this is completed, it will be assigned to the Pacific Fleet, and it is expected that VF-191 and VF-194 will be assigned to this carrier. These two squadrons are presently in the process of transitioning to the F-14 Tomcat.

OFFICIAL F-14 TACTICAL PAINT SCHEME

NAVY

35237

36320

36375

MIL-STD-2161(AS)

VF-1 WOLFPACK

*F-14A, 158984, was photographed at NAS Atlanta, Georgia, as it arrived for static display at the open house in June 1974. The aircraft was in the gull gray over white scheme. All squadron markings were red, including the **NK** tail code on the inside of the tail. VF-1 was assigned to the **USS Enterprise** at that time. Notice the AIM-54A Phoenix that was carried under the aircraft. The blue vertical stripes on the Phoenix indicate that it was a training missile. (Flightleader)*

VF-1 was commissioned for the seventh time on October 14, 1972. Initially the crews received transition training in the F-14 with VF-124, but with the arrival of the first Tomcat for VF-1, they separated from VF-124 on July 1, 1973. They were assigned to Attack Carrier Air Wing Fourteen (CVW-14). They had received their full complement of twelve F-14s by March 1974, and they deployed aboard the **USS Enterprise** on September 17, 1974. This was the first F-14 deployment, and was the only use of the Tomcat in Vietnam. During Operation FREQUENT WIND, which was the support of the final evacuation of American personnel from Vietnam, VF-1 compiled over 1400 flight hours and recorded over eight hundred arrested landings. They returned to NAS Miramar in May 1975.

The second deployment of the "Wolfpack" was also aboard **Enterprise,** and began on July 30, 1976. During this extended deployment, VF-1 logged 2150 hours and made 1035 arrested landings before returning to NAS Miramar in March 1977.

The third deployment was again on **Enterprise,** and lasted from April to October 1978. They participated in several fleet exercises in both the western Pacific and Indian oceans. A total of over 1900 flight hours and 938 arrested landings were recorded.

In 1980, VF-1 made its first deployment as part of CVW-2 aboard the **USS Ranger.** This was followed by a second tour aboard **Ranger** in 1982, during which time VF-1 logged its 17,000th accident-free flight hour that spanned five years of flying the Tomcat. They won the Safety **S** and the Clifton Award that year, signifying them as the best fighter squadron in the Navy.

In January 1984, while still part of CVW-2, VF-1 deployed to the western Pacific aboard the **USS Kitty Hawk.** The tour was an extended one, lasting until August of that year. Since then, the **Ranger** has returned to sea, and at present, VF-1 is assigned to CVW-2 and the **Ranger.**

This left side view of F-14A, 159465, was taken at Offutt AFB, Nebraska, in May 1977. This F-14A did not have an IR/TV pod under the nose, so only the ALQ-100 antenna was present. The aircraft was in the gull gray over white scheme with red unit markings. (J. Geer)

VF-1's CAG aircraft, 158979, was painted in a splinter camouflage pattern when this photograph was taken on May 22, 1977. This F-14 was painted in a dark, medium, and light gray pattern. All markings were in black, except for the red ejection seat warnings and the yellow ***RESCUE*** *markings.* *(Flightleader)*

A close-up of the tail illustrates the three grays that were on the aircraft. Notice the absence of the usual ***NK*** *tail code.* *(Flightleader)*

This elevated right front view gives us a look at the upper surfaces of 158979. Notice that the camouflage pattern extended on to the radome. *(Roth)*

A right rear view of the CAG's Tomcat provides a better look at the camouflage pattern applied to the lower portion of the aircraft. *(Grove)*

This left profile of 158979 shows more of the pattern of the three grays that was painted on this aircraft. The photograph was taken at NAS Miramar in May 1976. *(Kasulka via Geer)*

When this photograph of F-14A, 158981, was taken in December 1977, VF-1 had aircraft painted in the overall glossy gray scheme. The colorful unit markings were still in use as evidenced by this photograph. (Flightleader)

Another F-14A from VF-1 is pictured in the overall glossy gray scheme. This photograph was taken at NAS Miramar on October 29, 1977. The aircraft was armed with training Sparrow, Sidewinder, and Phoenix missiles for this open house display. (Bergagnini)

Right: F-14A, 159841, was photographed as it taxied out for a mission in January 1980. Notice the reduction in the red unit markings from that seen in the photograph above. The only red unit markings on this aircraft are on the tail and the ventral fin. Other markings were black, except for the colorful national insignia, the red ejection seat warnings, and the yellow **RESCUE** markings. The ACMI pod under the right wing was dark blue with an OD nose. (Grove)

Another change in markings is illustrated by F-14A, 159835, that was photographed at Dobbins AFB, Georgia, on June 29, 1980. The aircraft was an overall glossy gray with dark gray markings, to include the national insignia. The wolf's head and the tail code were red, and were smaller in size. Notice that the tail code had been changed to **NE** to represent the unit's change in air wing assignment and move to the **USS Ranger.** (Flightleader)

*All of the red had been replaced by gray when this photograph of F-14A, 161284, was taken at Luke AFB, Arizona, on September 3, 1984. The **NE** tail code is on the insides of the tails of this aircraft, and the wolf's head is on the outside of each tail. All markings are gray, except for the black **NAVY** on the rear fuselage, the red ejection seat warnings, and the yellow **RESCUE** markings.* *(Rogers via Cockle)*

*The tail of this F-14A illustrates another change that had occurred when this aircraft was photographed at NAF Andrews, Maryland, on November 10, 1984. The **NE** tail code and the wolf's head were on the outside of the tail in dark gray. Notice the different style of tail code from the one seen in the photograph above. The carrier name, **USS RANGER,** was on the lower portion of the tail. All other markings were gray, except for the black **NAVY** on the rear of the fuselage.* *(McGarry via Cockle)*

*VF-1 had been assigned to a different carrier when this photograph was taken in June 1983. Even though the **NE** tail code was still carried, the name **USS KITTY HAWK** was carried on the tail. The aircraft was overall glossy gray with dark gray markings, except for the black **NAVY** on the fuselage.* *(Grove)*

F-14A, 161292, was photographed in the experimental diagonal splinter camouflage scheme as it taxied out for a flight. This scheme was intended to hide the largest portion of the aircraft under various lighting and background conditions, and was applied in a diagonal pattern across the aircraft. Three shades of gray were used in the scheme seen here. All markings were in a contrasting shade of gray, and a blue training Sidewinder missile was carried on the glove pylon. *(Grove)*

VF-2 BOUNTY HUNTERS

*When this photograph of F-14A, 158987, was taken, VF-2 carried some of the most colorful markings applied to the F-14s. The aircraft was in the gull gray over white scheme with standard Navy markings. The **NK** tail code was black, shadowed with yellow. The fin cap was yellow, and the rudder was very dark blue with white stars. The ventral fin was yellow and dark blue with white stars. The large national insignia was carried on the nose, and was set on a background of red, white, and dark blue diagonal stripes. The **201** modex on the nose was black, shadowed in yellow.*
(Flightleader)

VF-2 undoubtedly has one of the most interesting and significant histories in all of Naval Aviation. It was first formed on July 1, 1922, and had the distinction of being the first carrier squadron deployed on the **USS Langley, CV-1.**

VF-2 was redesignated VF-6 on January 1, 1927, and a new VF-2 was formed on the same date. This was a squadron of enlisted pilots as the Navy tested the feasibility of using enlisted men as fighter pilots. The squadron became known as "The Chiefs' Squadron." But by the start of World War II, VF-2 had lost its identity as "The Chiefs' Squadron" since the enlisted pilots had been transferred to instructor positions ashore. The squadron was deployed aboard the **USS Lexington, CV-2,** when the carrier was lost at Coral Sea. The squadron was disestablished on July 1, 1942.

It was less than a year later on June 2, 1943, that the third VF-2 was established at Quonset Point, Rhode Island. In October of that same year, they deployed on board **USS Enterprise, CV-6,** flying the F6F Hellcat. They returned to Pearl Harbor with **Enterprise** in December 1943. Only three months later, in March 1944, VF-2 deployed aboard the **USS Hornet, CV-12.** This deployment was filled with action, with July 19th and 24th being particularly noteworthy. On those two days, VF-2 destroyed 118 enemy aircraft (51 and 67 respectively), with the loss of only one of their own Hellcats. Their World War II scoreboard is simply astonishing. The unit shot down 261 enemy aircraft, while destroying another 245 on the ground, for a total of 506. They destroyed 50,000 tons of enemy shipping, and flew 14,090 combat hours that included 2050 sorties and 184 strikes. Their own losses included three aircraft to aerial combat and four lost to anti-aircraft fire. At the end of the war, VF-2 was again disestablished on November 9, 1945.

Twenty-seven years later, on October 15, 1972, VF-2 was commissioned for the fourth time. The name "Bounty Hunters" was selected for the unit, along with the tactical call sign "Bullet." The original insignia, that had been painted on the Curtis F6Cs that comprised the first squadron to ever deploy aboard a carrier, was selected for the new F-14 unit. The stripe of red, white, and blue diagonal bands that adorned VF-2's first Tomcats in the days of the colorful paint schemes, was known as the "Langley Stripe," and was used on the **Langley's** first aircraft.

The first F-14 deployment for VF-2 was with VF-1 aboard the **USS Enterprise, CV(N)-65,** as the Tomcat made its first carrier deployment. Like VF-1, the "Bounty Hunters" participated in Operation FREQUENT WIND, which was the successful evacuation of Saigon. Later, they became the first F-14 squadron to win the Safety "S" award.

After three cruises aboard **Enterprise,** VF-2 was reassigned to CVW-2, and went to WESTPAC for the fourth time aboard the **USS Ranger, CV-61.** A fifth deployment followed, again aboard **Ranger,** and a sixth was made while embarked on the **USS Kitty Hawk, CV-63.** In 1985 VF-2 surpassed the 10,000 hour accident free mark as they prepared for their 1986 cruise aboard **Ranger.** The squadron remains assigned to CVW-2 and the **Ranger** as of July 1987.

*The left side view of F-14A, 158998, shows some experimental unit markings on the tail of this aircraft. Compare these markings with those on the aircraft visible behind the nose. The **NK** tail code was carried on the inside of the tail in black. The ventral fin had three small diagonal red, white, and dark blue stripes instead of the usual markings seen there. All other markings were standard for VF-2.* *(Roth)*

*At left is a right front view of VF-2's CAG, 158985, in a splinter camouflage pattern. The aircraft was in three shades of gray with black markings. At right is an elevated left front view of the same aircraft. The pattern on the upper surfaces is easy to see here. The national insignia on the nose and the **NAVY** on the rear fuselage were gray.*
(Left Logan, right Flightleader)

At left is a right rear view of the CAG aircraft that shows the camouflage pattern very well. At right is a left rear view of the aircraft that illustrates the camouflage pattern on that side. Notice the small red, white, and dark blue diagonal stripes that were applied to the ventral fin. *(Both Flightleader)*

F-14A, 159843, of VF-2 was seen at Offutt AFB, on May 10, 1981. Notice that the carrier assignment has changed again as indicated by **USS RANGER** under the **NE** tail code. The tail code was black, shadowed in yellow. The rest of the markings were the same that VF-2 had carried on the **USS Enterprise.** (Cockle)

At left is a close-up of the tail of 159843 which shows the markings more clearly. The fin cap was still yellow, although it is hard to see in this photograph. At right is a shot of the nose of the same aircraft which gives an excellent view of the markings that were applied there. (Both Cockle)

The photograph at left provides a right front view of 161273 that was taken in June 1983. Notice the change in the carrier name in the view at right. **USS KITTY HAWK** was painted in black under the tail code. (Both Grove)

Above: Another step toward the low visibility markings is seen in this photograph of F-14A, 161291, that is dated February 1986. VF-2 had changed their markings to a white skull on a black circle which was outlined in yellow. The badge is superimposed over three diagonal stripes of red, white, and dark blue, outlined with yellow. The aircraft was overall glossy gray with black markings. *(Grove)*

Left: F-14A, 161295, was photographed in subdued markings on April 27, 1985. The aircraft was overall glossy gray with black markings. The carrier name, ***USS RANGER,*** *was painted behind the glove vane.* *(Kaston)*

The markings were barely visible on F-14A, 159837, when this photograph was taken in February 1986. The aircraft was in the overall gray scheme, with the markings being in a slightly darker shade of gray. The modex ***214*** *on the nose was a dark gray, and is about the only marking that can easily be seen.* *(M. Geer)*

A VF-2 Tomcat was taxiing out for a training mission when it was photographed in February 1986. The aircraft was in the tactical blue/gray scheme with dark gray markings. The only color on the aircraft was the blue on the training Sidewinder missiles. *(Grove)*

F-14A, 159869, made an appearance at MCAS El Centro, California, on March 24, 1985. The aircraft was painted in the tactical blue/gray scheme, and had gray markings. The radome appears to be a different shade of gray, possibly from another aircraft. *(Huston)*

VF-21 FREELANCERS

*F-14A, 161613, of VF-21, was photographed in full color markings in September 1984. The aircraft was in the overall glossy gray scheme with black markings. Notice the black that extends from the radome to behind the canopy. The crew names were yellow on black canopy rails. On the tail was the squadron emblem in black. It was on a yellow chevron which was outlined in black. The stylized **NK** tail code was carried on the inside of the tail. The fin cap was yellow with a black line under it. **USS CONSTELLATION** was below the squadron markings on the tail. The ventral fin had a black flash, and the **VF-21** carried there was black, shadowed in yellow. The **205** modex on the nose was black, shadowed in yellow.* *(Grove)*

The "Freelancers" of VF-21 trace their history back to March 1, 1944, when Fighter Squadron Eighty-One was formed at NAS Atlantic City, New Jersey. After flying the F6F Hellcat in major campaigns during World War II, the unit went through successive redesignations of VF-13A, VF-131A, VF-64, and on July 1, 1959, the present designation of VF-21. During this time they flew the F8F Bearcat, F4U Corsair, F9F Panther, F2H Banshee, F3H Demon, and the F4B, -N, -J, and -S versions of the Phantom, before receiving their present F-14 Tomcats. On June 17, 1965, while flying the F-4B, two VF-21 Phantoms shot down two MiG-17s, thus scoring the first MiG kills of the Vietnam War.

In 1983, VF-21 retired its Phantoms, and this ceremony was attended by "Phantom Phlyers and Phixers" of all ages, because it marked the last appearance of the F-4 in active naval service in the continental United States. In March of the following year, the squadron became active with its new F-14 Tomcats, and became part of CVW-14 assigned to the **USS Constellation.** Their first deployment with the F-14 began in early 1985, and the squadron remains a part of CVW-14 today.

*At left is the CAG aircraft of VF-21, 161603, as it appeared on May 5, 1984. There were no special CAG markings except for the **200** modex on the nose and the **00** at the top of the rudder. Both of these were black, shadowed with yellow. At right is F-14A, 161617, as it taxied out for a training mission in September 1984. The aircraft carried a blue ACMI pod on the glove pylon.* *(Both Grove)*

shows the markings on the lower portion of the fuselage more clearly. The aircraft was on May 5, 1984. *(McGarry via Cockle)*

of VF-21's CAG aircraft. ***CAPT. SWEDE ZERR,*** *appeared on the canopy rail in yellow.* ***THE YEAR*** *under the RIO's name. At right is a view of the rear of the aircraft which lized* ***NK*** *tail code.* *(Both McGarry via Cockle)*

F-14A, 161607, was photographed in June 1986, as it taxied out for a flight. The ***NK*** *tail code has been moved to the rudder on the outside of the tail, and is now vertical. The style of the letters has also been changed.* *(Grove)*

VF-24 FIGHTING RENEGADES

A VF-24 F-14A, 159629, is seen here in June 1978. The aircraft was in the gull gray over white scheme. The modex ***203*** *on the nose, and the* ***03*** *on the rudder, were black, shadowed in red. The squadron markings on the tail consisted of a red check, highlighted with black. The* ***NG*** *tail code was black, shadowed in white. The white ventral fin had a red check, outlined with black. All other markings were standard. Part of the carrier name,* ***USS CONSTELLATION,*** *is visible on the inside of the tail.*
(Grove)

Fighter Squadron 24 was originally commissioned as Fighter Squadron 211 in June 1955. It deployed aboard the **USS Bon Homme Richard, CVA-31,** in August 1956, while flying the FJ-3 Fury. In 1958, it embarked aboard the **USS Midway, CVA-41** with its new F8U Crusaders. It was redesignated VF-24 the following year on March 9, 1959. The next change in carrier assignment was in November 1965 when the squadron deployed aboard the **USS Hancock, CVA-19.**

In January 1967, VF-24 was back aboard the **USS Bon Homme Richard** for its third combat tour. On May 19, 1967, two VF-24 pilots downed the squadron's first two MiGs with Sidewinder missiles. Two more MiGs were destroyed and a third one was counted as a "probable" on July 21. One of the kills was scored with the 20mm cannon, and the other was achieved with a Sidewinder. Two more combat tours were made aboard **Hancock,** one in 1973 and one in 1975.

The Tomcat replaced the Crusader when VF-24 received its first F-14 on December 9, 1975. By March 1, 1976, the squadron became part of Carrier Air Wing Nine, and remains a part of that air wing to this day. In March 1977, VF-24 became the first F-14 squadron to win the coveted "MUTHA" award, which is presented annually to the best Pacific Fleet fighter squadron. The first F-14 carrier deployment began the following month aboard the **USS Constellation, CV-64.** The second consecutive "MUTHA" award was presented in March 1978. This was followed in July by the Clifton Award which recognizes the best fighter squadron in the Navy. In August, the squadron began its second deployment to WESTPAC aboard the **Constellation.**

On April 18, 1980, the "Renegades" embarked again aboard the **Constellation,** and this deployment lasted until August 5, 1980. This was the longest at-sea period ever by a west coast carrier. In 1982 VF-24 and Carrier Air Wing Nine were reassigned to the **USS Ranger,** but after that carrier suffered a major fire in November of 1983, the squadron returned to NAS Miramar in February 1984. They are now assigned to the **USS Kitty Hawk,** along with the rest of CVW-9.

This left side view of F-14A, 159613, shows the markings that VF-24 carried. The radome was cream and white.
(Swanberg via Geer)

*VF-24's CAG aircraft, 159631, was photographed in June 1976. Above the **NG** tail code was a black **9,** surrounded by multi-colored stars. This was to represent Carrier Air Wing 9.* *(Flightleader)*

Center: A right front view of the CAG aircraft is seen in this May 1976 photograph. It was taken at NAS Miramar. *(Kasulka via Geer)*

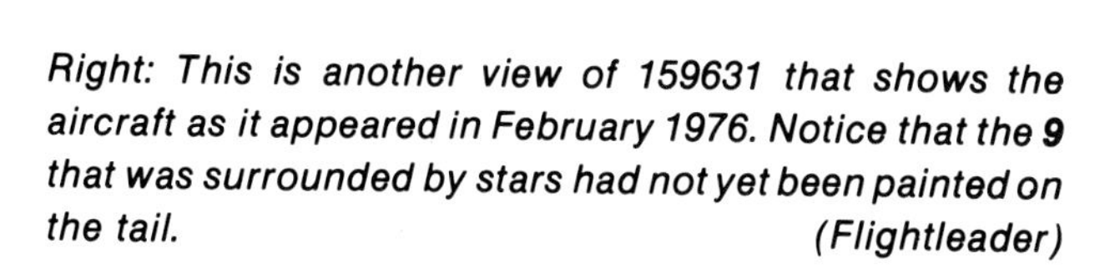
*Right: This is another view of 159631 that shows the aircraft as it appeared in February 1976. Notice that the **9** that was surrounded by stars had not yet been painted on the tail.* *(Flightleader)*

At left is an overall view of F-14A, 159629, that was taken on May 6, 1978. The squadron markings on the tail appear weathered, and the red has worn off at several points. At right is a close-up of the three drone kills that were carried on the nose of this Tomcat. The drone kills were painted red. (Both Knowles)

F-14A, 159463, was on the transit line at Dobbins AFB, Georgia, on June 4, 1978. The aircraft was still in the gull gray over white scheme, however, the colorful squadron markings have disappeared. The ***NG*** *tail code had also been changed to a different style from that carried on the aircraft at the top of this page. (Flightleader)*

Here is a left side view of a VF-24 Tomcat as it appeared on May 6, 1978. This photograph illustrates the loss of the colorful markings. (Knowles)

Left: F-14A, 159468, photographed at NAS Miramar on January 29, 1978, reflects the toned down paint schemes that were starting to make their appearance about that time. Notice that the aircraft was still in the gull gray over white scheme. (Flightleader)

*VF-24's CAG, 159458, is seen here as it appeared in the new low visibility scheme on June 17, 1978. The aircraft was overall glossy gray, and had dark gray markings. Note the return of the squadron marking in dark gray on the tail below the **NG** tail code. The carrier name, **USS CONSTELLATION,** appeared on the inside of the tail. (Flightleader)*

Center: This top view of F-14A, 160887, was taken at Offutt AFB as it taxied out for take off. The large colorful national insignia was still being carried on the front of the fuselage, even though the aircraft was painted in the low-visibility scheme. (Cockle)

Right: In 1978, VF-24's Tomcats were usually seen in the overall gray scheme, with dark gray markings, as illustrated in this photograph taken in September of that year. (Kasulka via J. Geer)

*One of the more interesting color schemes to appear was the one seen on VF-24's CAG, 159593, as photographed in June 1981. The aircraft was overall glossy gray, with black vertical tail surfaces. The **NG** tail code, the squadron markings, and the CAG markings on the tail were all red. The ventral fin was black with a red outline, and it had **VF-24** on it in red. An **S** in black, shadowed in red, appeared on the nose to represent the safety award.* *(Grove)*

Center left: Another style of markings for VF-24 was seen on F-14A, 159627, that was photographed on the transit ramp at Dobbins AFB, on June 5, 1982. The aircraft was overall glossy gray, with dark gray markings. The squadron markings on the tail were dark gray, shadowed in red. The ventral fin had a dark gray and red flash on it. Notice the small, colorful national insignia carried on the front of the fuselage. *(Flightleader)*

Center right: The same aircraft seen at center left is seen here with another change toward low visibility markings. The red shadowing has been removed from the squadron markings on the tail and the ventral fin. The national insignia is now also in gray. *(Grove)*

Left: A close-up of the tail of 159635 illustrates the squadron markings that had the red shadowing. The dark gray and red on the ventral fin is also visible. *(Cockle)*

The same aircraft that is at the top of page 22 is seen here in another variation of the CAG markings. The aircraft was photographed at NAS Miramar on May 5, 1984, and was in the overall gray paint scheme with dark gray markings. The CAG markings on the tail were all dark gray, and all of the colorful stars had been repainted dark gray. The style of the squadron marking had been changed from what was seen on the previous page. The only color on this aircraft, which doesn't show in this photograph, is the red showing on the upper portion of the squadron markings on the tail.
(McGarry via Cockle)

Center: The right side of F-14A, 159619, shows the markings that were carried there. Notice that no carrier name was painted on the aircraft at the time this photo was taken. *(McGarry via Cockle)*

Right: This close-up of the tail of 159627 was taken at NAS Fallon, Nevada, on March 8, 1983. Notice the small Playboy bunny in light gray on the squadron markings. All markings were dark gray over the glossy gray paint scheme. *(Grove)*

Above: VF-24 had started using the tactical blue/gray scheme by the time this photograph of F-14A, 160887, was taken on September 8, 1984. The aircraft was two tones of gray with dark gray markings. *(Flightleader)*

Left: A close-up of the tail of 160887 shows details of the markings that were painted there. Most of these markings are not clearly visible in the photograph above. Notice the carrier name, ***USS RANGER****, above the* ***NAVY.*** *(Flightleader)*

The same F-14A as seen above was photographed at NAF Andrews on June 7, 1986. The modex had been changed from ***212*** *to* ***210*** *when this photograph was taken. All other markings remained the same.* *(McGarry via Cockle)*

VF-51 SCREAMING EAGLES

*F-14A, 160685, of VF-51, was photographed on board the **USS Kitty Hawk** on April 28, 1979. The aircraft was overall glossy gray with black markings. The vertical tail was gloss black with red horizontal stripes. The **NL** tail code was black, outlined in white.*
(Huston)

Fighter Squadron Fifty-One has the distinction of being the oldest fighter squadron of continuous service in the Pacific Fleet. Originally commissioned as VF-3S "Striking Eagles," they have been through several designation changes before becoming the present "Screaming Eagles" of VF-51.

The "Screaming Eagle" insignia first appeared on a Curtis F6C-4 in 1927. In the pre-World War II years, the squadron flew the Boeing FB-5, F3B-1, F4B-4, F2F-1, and F3F-3. At the outbreak of the war, the squadron was operational in the F4F-3. After flying their Wildcats at Guadalcanal, they introduced the F6F Hellcat into fleet service. In February 1945, they transitioned to the F4U Corsair. They were deployed aboard the **USS Franklin** when it was severely damaged by kamikazes. But after the war, the squadron continued flying Grumman's products when they transitioned to the F8F Bearcat.

On November 18, 1948, VF-51 said goodbye to the Grummans and to props as it entered the jet age in the straight-winged FJ-1 Fury. They became the first fleet squadron to operate jets aboard a carrier in March 1948. Later they became the first to take jets into combat. They made the first Navy air-to-air kills when two of their F9F Panthers shot down two Yak-9s on July 3, 1950. From the Panthers, they moved on to the swept wing Cougars, then to the FJ-3 Fury in November 1954. They operated these aircraft until exchanging them for the supersonic F11F Tiger, an aircraft that they flew for only a short time before transitioning to the F4D-1 Skyray. The Skyray lasted only two years with VF-51, making one cruise aboard the **USS Ticonderoga, CVA-14.**

In November 1960, the "Screaming Eagles" again changed aircraft, this time acquiring the F8U-1 Crusader. These were upgraded to F-8Es in 1962. After evaluating the air-to-ground rocket and bombing capability of the Crusader, VF-51 flew secret missions into Laos beginning in June 1964. Still deployed aboard the **Ticonderoga,** the squadron flew in support of the **USS Maddox** when it came under fire from North Vietnamese torpedo boats in August 1964. They flew the first retaliatory strike against North Vietnam on August 5.

In 1968, while flying the F-8H, two MiG-21s were shot down by two of VF-51's pilots. The following year, they became the first squadron to ever win the Clifton Trophy as the Navy's most outstanding fighter squadron.

In 1971 VF-51 transitioned to the F-4 Phantom, and were reassigned to CVW-15. They returned to SEA in their new aircraft, and shot down four MiG-17s. In 1975, they flew in support of the Vietnam evacuation, and later were involved in the **USS Mayaguez** rescue.

Transition to the F-14 Tomcat began in late 1977, and they made their first F-14 deployment aboard the **USS Kitty Hawk** in May 1979. A second deployment followed in April 1981. In January 1983, they flew across the country to join their new carrier, the **USS Carl Vinson, CV(N)-70.** They deployed in March, and completed a round-the-world cruise, returning to NAS Miramar.

During their following deployment, they became the first squadron to do F-14 day and night automatic carrier landings, and they performed the first intercepts of a Soviet "Backfire" bomber, as well as "Floggers" and "Flagon" fighters. A six month cruise was made aboard the **Vinson** in 1986. Today the squadron is part of CVW-15, and remains aboard the **Vinson.**

VF-51's CAG aircraft is the subject of this photograph, and is shown as it appeared in October 1982. The only change from the markings seen on the previous page is the smaller, gray national insignia applied to the front of the fuselage. (Grove)

A close-up of 160655 gives a good look at the CAG markings on the tail. The CAG stripes were applied to both sides of the vertical tail, and were, from top to bottom, red, green, orange, blue, and yellow. (Flightleader)

This detail shot of the nose of the CAG aircraft shows the squadron badge that was displayed under the canopy. The badge was an eagle over a yellow star, set on a dark blue background. The badge was surrounded by a white circle, outlined with black. ***VF-51*** *appeared in black on a white ribbon below the badge. (Flightleader)*

The complete right side of the CAG aircraft is seen in this photograph that was taken on May 5, 1984. Notice that the squadron badge was not carried under the canopy rail at that time. (McGarry via Cockle)

F-14A, 159837, is seen here as it appeared at NAS Miramar on May 5, 1984. The squadron was still using full color markings on its aircraft at that time. *(McGarry via Cockle)*

This close-up view of the tail of a VF-51 aircraft illustrates the red squadron markings that were applied to both sides of the vertical tail surfaces. The glossy black paint shows up very well here. ***(Flightleader)***

As late as 1985, VF-51 still used colorful markings on their aircraft, as evidenced by this photograph taken at McGuire AFB. The carrier name, USS KITTY HAWK, was behind the glove vane in black. ***(Spering via J. Geer)***

VF-51's CAG, now painted in subdued markings, was photographed as it taxied out for a training mission. The aircraft was overall glossy gray with black markings. Three red horizontal stripes appeared on the tail under the ***NL*** *tail code. The code was black, shadowed in white. The crew names were white on the black canopy rail.* *(Grove)*

Another F-14A from VF-51 is shown taxiing at NAS Fallon. The aircraft was in the tactical blue/gray scheme, but had a temporary two tone gray camouflage applied over the standard paint scheme. Notice the red squadron markings on the tail that were barely visible under the additional camouflage. *(Grove)*

*This F-14A waits for permission to taxi out in March 1984. The aircraft was in tactical blue/gray with black modex **110,** crew names, and **USS CARL VINSON.** All other markings were a dark gray except for the squadron markings on the tail, the **NL** tail code, the **10** at the top of the rudder, and the **VF-51** on the ventral fin. These were all in white.* *(Grove)*

A true tactical blue/gray scheme was carried on this VF-51 Tomcat that was seen at NAS Miramar in December 1982. The aircraft has contrasting light and dark gray markings. *(Grove)*

*Another variation of the tactical scheme was photographed at NAF Andrews on December 28, 1985. Notice the difference in the **NL** tail code on this aircraft as compared to that seen in the photograph at center right of this page. The modex **106** on the nose and **06** on the top of the rudder were a dark charcoal gray.* *(McGarry via Cockle)*

VF-111 SUNDOWNERS

This photograph shows F-14A, 160686, of VF-111, as it appeared at NAS Miramar on March 22, 1979. The aircraft was in the overall gray scheme with red and white squadron markings. The carrier name, ***USS KITTY HAWK,*** *was painted in red behind the glove vane. The rising sun markings were applied to the ventral fin in red and white, with* ***VF-111*** *in white on the rear of the fin.*
(Huston)

The forerunner to the present VF-111 was commissioned in San Diego on October 10, 1942, as Fighter Squadron Eleven. Two weeks later they were in Hawaii training in their F4F-4 Wildcat fighters. Only three months later they were on their way to the combat zone. They began working from flight strip #2 on Guadalcanal, flying intercept and escort missions with Torpedo Eleven and Bombing Eleven. From April to July 1943, they accounted for fifty-six confirmed kills. For so effectively downing the rising sun symbol of Japan, they earned the nickname that they still carry today, the "Sundowners."

Their second combat tour began in October 1944 aboard the **USS Hornet, CV-12.** They had transitioned to the F6F Hellcat in the interim, and flew strikes against Formosa, Luzon, and the Manila airfields. They also were involved in the battle of Leyte Gulf.

After World War II, the "Sundowners" were home-ported in San Diego, and redesignated VF-11A. They transitioned to the F8F Bearcat, but retained four F6F-5Ps for photo work. In July 1948, they received their present designation of VF-111.

After transitioning to the F9F-2 Panther, they entered the Korean conflict flying from the deck of the **USS Philippine Sea, CV-47.** On November 9, 1950, LCDR W. T. Amen, the squadron's commanding officer, shot down a MiG-15, scoring the first Navy jet kill in aerial combat. Also during this tour, LT Carl E. Dace had the dubious distinction of being the first Navy pilot to use the ejection seat in combat. A second combat tour was made aboard the **USS Boxer, CV-21,** and a third was made aboard the **USS Lake Champlain, CV-33.** During the last combat deployment, the squadron made the last Navy strike of the war on July 27, 1953, the day the truce was signed.

After the Korean combat tours, VF-111 made many deployments to the western Pacific, flying varied aircraft that included the F9F-2 Panther, F9F-6 Cougar, FJ-3 Fury, F11F Tiger, and the F8U Crusader. Deployments were made aboard several carriers including **Wasp, Lexington, Hancock, Kitty Hawk, Midway, Bennington, Oriskany, Ticonderoga,** and **Shangri-La.** On January 19, 1959, VF-111 was decommissioned, but the following day VA-156 was redesignated VF-111, thus continuing the "Sundowner" tradition.

During the war in Vietnam, VF-111 again made combat cruises flying the F-8 Crusader. They flew over 12,500 missions in SEA. In the 1967-68 time frame, a detachment from VF-111, known as "Omar's Orphans" was assigned to the **USS Intrepid, CVA-11,** to fly escort for the attack and recon aircraft embarked on that carrier. During this tour, LT Tony Nargi shot down a MiG-21. The squadron followed this deployment with a combat tour aboard the **USS Ticongeroga, CVA-14,** in 1969.

In 1971, the squadron was reassigned to CVW-15, and began transitioning to the F-4B Phantom. In November, they embarked aboard the **USS Coral Sea, CVA-43,** for their last combat tour. After the end of hostilities, the squadron remained a part of CVW-15, and made several more tours aboard the **Coral Sea.** They transitioned to the F-4N in 1974.

In June 1976, the squadron was reassigned to CVW-19 and the **USS Franklin D. Roosevelt, CVA-42.** They made a Mediterranean deployment aboard the **Roosevelt,** completing 15,800 accident free flight hours. They returned to the west coast and NAS Miramar in April 1977. By the end of that year, they were well into their transition to the F-14 Tomcat. They officially "stood up" in F-14s on October 1, 1978. Their first deployment with the Tomcat was aboard the **USS Kitty Hawk, CV-63,** beginning on May 30, 1979. A second tour followed, again on the **Kitty Hawk,** in 1981.

While remaining a part of CVW-15, VF-111 was transferred to the **USS Carl Vinson, CV-70,** and like their sister squadron in the wing, VF-51, embarked on the new carrier for a round-the-world cruise that began at Norfolk, Virginia, on March 1, 1983. It ended at Alameda on October 29. Today, VF-111 remains a part of CVW-15 assigned to the **Vinson.**

*At left is a front view of 203, which provides a good look at the sharksmouth that was carried there. The **203** modex was red, outlined with white. The sharksmouth was also red and white. The eye above the sharksmouth was gray, as was the national insignia on the fuselage. Notice the TARPS pod that was carried under the aircraft. At right is F-14A, 160676, that was photographed on April 24, 1980. Notice the different style of sharksmouth from that on the aircraft at left. The eye above the sharksmouth was also different, being red and white instead of gray. This aircraft still carried the large, colorful national insignia on the fuselage.* *(Left Linn via Cockle, right Cockle)*

*Above: This overall view of 160676 shows the full markings that were applied to the aircraft. The fin caps were red, while the **NL** tail code was red, shadowed in white.* *(Cockle)*

Left: F-14A, 160680, was a visitor to the Davis-Monthan AFB, Arizona, transit ramp during October 1980. The aircraft was in full color markings for VF-111. *(Rogers via J. Geer)*

VF-111 was moving toward the low visibility style of markings when these photographs were taken in June 1980. At left is the right side of 160674, which was in the overall glossy gray scheme with dark gray markings. The squadron sunburst markings had been moved from the ventral fin to the vertical tail. The sharksmouth was also painted a dark gray, as was the large national insignia on the fuselage. Notice that the carrier name was painted on the bottom of the intake instead of behind the glove vane. At right is the other side of the aircraft, providing an excellent view of the markings. *(Both Grove)*

Above: VF-111's CAG aircraft, 160666, is seen as it appeared on January 25, 1981, in full color markings. The carrier name, ***USS KITTY HAWK,*** *was in red on the bottom of the intake. The* ***NL*** *tail code was red, shadowed with white, as was the* ***00*** *on the top of the rudder. The* ***200*** *modex on the nose was red, outlined in white.*

(Huston)

Right: This close-up gives a good view of the CAG markings that were applied to the tail. The various aircraft silhouettes of the air wing were painted down the rudder in different colors. *(Grove)*

By mid-1981, the markings on the CAG aircraft had been changed. In the photograph at left, 160656 is seen in the new markings at Offutt AFB. At right is a close-up of the tail, which exhibits the sunburst that was painted there. Notice that the sun is at the center of the bottom of the rudder instead of the front corner, as seen on the aircraft at the bottom of the previous page. The sun's rays are outlined in black, red, yellow, black, blue, and green, from left to right.

(Both Cockle)

*Another change in markings is illustrated by this photograph of F-14A, 160664, that was taken in February 1982. The sunburst is on the tail, with the sun at the bottom right of the tail. The **NL** tail code was red, shadowed with white, on the inside of the tail. **VF-111** was in red on the ventral fin, as was the modex **213** on the aileron.* *(Grove)*

*Still another change to the tail of a VF-111 aircraft is seen in this view of F-14A, 158984. The sun is now centered on the bottom of the tail, and the **NL** tail code is on the outside of the tail in red, shadowed with white. The white portion of the sunburst is outlined with yellow. The ventral fin is red with a yellow outline, and **VF-111** is in yellow.* *(Grove)*

*VF-111's CAG aircraft has taken a step toward low visibility markings in this photograph. The aircraft is overall gray with black markings. The red squadron markings have been moved to the rudders, and there is no white on them. The carrier name, **USS CARL VINSON,** is on the bottom of the tail in black.* *(Grove)*

*This F-14A is completely devoid of any color in its markings. Instead, it is overall glossy gray with dark gray markings. The **NAVY** on the rear fuselage, the **NL** tail code, and the **06** on the rudder are black.* *(Cockle)*

Above: VF-111's wide variety of markings included this aircraft in overall glossy gray with gray markings. The bureau number, ***NAVY*** *on the fuselage, and the* ***VF-111*** *on the ventral fin were black.* *(Huston)*

Right: F-14A, 160678, in the tactical blue/gray scheme, awaits clearance to taxi out for a mission in October 1982. All markings were a contrasting gray over the tactical blue/gray scheme. *(Grove)*

At left is a view of 160668, with a different style of the tactical scheme than previously seen. The markings are a dark gray over the tactical blue/gray. At right is 161152, which has a tactical blue/gray scheme with gray and black markings. The ***NL*** *tail code,* ***12*** *at the top of the rudder,* ***USS CARL VINSON*** *on the tail, the* ***VF-111*** *on the ventral fin, the* ***212*** *modex on the nose, and the crew names were all black.* *(Both Grove)*

VF-114 AARDVARKS

A VF-114 F-14A, 159852, was photographed on climb-out in July 1978. The aircraft was in full squadron colors on the gray and white paint scheme. The orange stripe on the front of the fuselage was outlined in black.
(Picciani Aircraft Slides)

The roots of VF-114 extend all the way back to the commissioning of VB-5B on July 2, 1934, at Norfolk, Virginia. They moved to the west coast in 1935, and changed their designation to VB-2 in July 1937. They were flying the F4B-4 at that time, but replaced it with the SB2V-2 in March 1938.

Early in World War II, the squadron embarked aboard the **USS Lexington, CV-2,** flying the SBD Dauntless dive bomber. The carrier was lost at the battle of Coral Sea in May 1942, and VB-2 was decommissioned on July 1.

On October 10, 1942, VB-2's former commanding officer commissioned VB-11, which was equipped with the SB2C-1C Helldiver. While deployed aboard the **USS Hornet, CV-12,** the squadron shot down two enemy aircraft, destroyed another seventy-five on the ground, damaged thirteen warships, and sunk eight merchant ships. They participated in strikes against airfields, factories, docks, and harbor installations.

After the war, VB-11 was redesignated VA-11A on November 15, 1946, while based in Hawaii. They then returned to NAS San Diego later that month, and made a 1947 world cruise aboard the **USS Valley Forge.** In July 1948, they were redesignated VA-114, and then VF-114 on February 15, 1950. They then transitioned to the F4U Corsair, and deployed aboard the **USS Philippine Sea** on July 5, 1950. During the Korean Conflict, they flew more than 1100 strikes against the North Koreans and Chinese Communists.

Following Korea, the squadron entered the jet age, flying first the F9F Panther, then the F2H Banshee. The F2H was followed by another McDonnell product, the F3H Demon, in 1957. In 1961, yet another McDonnell fighter replaced the Demon, as VF-114 became the first Pacific fleet squadron to receive the F-4B Phantom. They were assigned to CVW-11, and made their first deployment in Phantoms in September 1962 aboard the **USS Kitty Hawk.** While flying the Phantom they made five combat cruises in SEA. They shot down five enemy aircraft and delivered hundreds of tons of ordnance against ground targets.

In January 1976, the "Aardvarks" transitioned to the F-14 Tomcat, and became operationally ready in the second half of that year. Their first cruise began in October 1977, with the squadron still being assigned to CVW-11 and the **Kitty Hawk.** In March 1979, they embarked aboard the **USS America, CV-66,** with the rest of CVW-11 for a cruise in the Mediterranean - an unusual event for a west coast squadron. A second Mediterranean tour was made aboard **America** in 1981, with the squadron returning to NAS Miramar in November of that year.

In September 1982, CVW-11 and VF-114 returned to WESTPAC aboard the **USS Enterprise, CV(N)-65.** Today the "Aardvarks" remain a part of CVW-11 which is assigned to the "Big E."

*This close-up shows the tail of 159852 as it appeared on May 22, 1977. The aircraft was the CAG aircraft at that time, as can be seen by the **00** modex on the top of the rudder. The aardvark, in orange on a black shadow, can be clearly seen in this photograph. The fin cap was orange.* *(Flightleader)*

A move toward low visibility can be seen in this photograph of a VF-114 F-14A, 159874, that was taken in May 1980. The aircraft was overall glossy gray, and still carried the orange stripe, outlined with black, on the front of the fuselage. Notice that the orange has disappeared from the ventral fin and the top of the tail. The carrier name, ***USS AMERICA,*** *was carried on the bottom of the tail in black.* *(Flightleader)*

Center: Another step toward low visibility markings is evident in this view of F-14A, 159833, that was photographed at NAS Miramar, on May 5, 1984. The aircraft was overall glossy gray and still carried the orange stripe, outlined in black, on the nose. The aardvark on the tail had been changed to all black, and the carrier name was ***USS ENTERPRISE,*** *which was painted in black under the* ***NH*** *tail code. All other markings were in gray, including the small national insignia located under the orange stripe on the nose.* *(McGarry via Cockle)*

Left: Color had returned to VF-114 by the time this photograph was taken in August 1986. The orange stripe had disappeared from the front of the aircraft, however, the aardvark was orange, standing on a black shadow. This aircraft also had a small national insignia in color, with all other markings being in black. There is a small aardvark painted on the front of the right external fuel tank. *(M. Geer)*

*F-14A, 159841, of VF-114, appeared in the new tactical blue/gray scheme in September 1983. The aircraft still carried the aardvark on the tail. It was painted in orange, and was standing on a black shadow. All other markings were black, except for the national insignia on the nose, and the **NAVY** on the rear fuselage, which were in contrasting gray.* *(Grove)*

A close-up of the tail of 159841 shows that the aardvark on the tail had been changed to black by the time this photograph was taken on March 26, 1987. *(Kinzey)*

Compare this shot of the tail of 159841 with the one at left to see the difference in the aardvark marking on the tail. The aardvark in this photograph was still orange, standing on a black shadow. *(Grove)*

*At left is the right side of F-14A, 159833, that was seen at NAS Miramar on March 26, 1987. The aircraft was overall tactical blue/gray, with black markings. At right is the left of the same aircraft as it appeared on May 5, 1984. Notice that a dark gray stripe, outlined in black, was carried on the front of the fuselage. All markings were black, except for the national insignia on the front of the fuselage, and **NAVY** on the rear fuselage, which were dark gray.* *(Left Kinzey, right Rogers)*

VF-124 GUNFIGHTERS

One of the most colorful F-14As, was painted in this Bi-centennial scheme by VF-124. The aircraft is seen as it appeared at NAS Miramar, on November 1, 1975. The aircraft was in the gray over white scheme with extensive use of red, white, and blue markings. *(Brewer)*

VF-124 is the west coast Fleet Readiness Squadron or Replacement Air Group (RAG) for the F-14 Tomcat. They were originally commissioned in April 1958 when swept wing jet fighters necessitated an expansion of the Navy's flight training program. They were initially tasked with several missions. First, they provided F-8 Crusader combat flight training for all fleet replacement pilots. Second, they were responsible for basic and refresher all-weather training for Pacific Fleet aviators, and they offered maintenance training for personnel in key enlisted ratings critical to fleet F-8 squadrons. They maintained combat-ready pilots and enlisted men for fleet operations in case of a national emergency, and provided fighter support of the Air Defense Command in conjunction with the Air Force. The unit continues this duty today.

On June 30, 1961, VF-124 moved from Moffet Field to NAS Miramar. It continued as the F-8 training squadron for over a decade, becoming known as the "Crusader College." They became the first F-14 training squadron in August 1972. It was a crew from VF-124 that took the Tomcat to the Paris Air Show in 1973 when the F-14 was first shown to the world in a series of impressive flight demonstrations. The first carrier qualifications for replacement pilots began in December 1974 aboard the **USS Kitty Hawk.** Training with the TARPS (Tactical Air Reconnaissance Pod System) began in 1980.

F-14A, 158629, of VF-124 illustrates typical markings of the mid-seventies. The aircraft was in the gray over white scheme with red stripes on the tail. The ***NJ*** *tail code was black, shadowed with white, as was the* ***410*** *modex on the nose. The large, colorful national insignia was carried on the front of the fuselage, with all other markings being in black.* *(J. Geer)*

This close-up of the tail of one of VF-124's Tomcats, shows the red stripes that were applied to the upper portion of the horizontal tail surfaces. Also notice the lack of white shadowing on the ***NJ*** *tail code as compared with the aircraft in the photograph to the left. The modex* ***27*** *is in black on the top of the rudder, as compared to the modex in the photograph to the left.* *(Brown)*

One of the last gray over white schemes used by VF-124 is seen in this line-up of aircraft at NAS Miramar on August 11, 1975. These aircraft were gray over white with black markings. Other than the colorful national insignia and the yellow ***RESCUE*** *markings on the front of the fuselage, all color had been removed.* *(J. Geer)*

At left is the right side of F-14A, 158622, in VF-124 markings. The aircraft was overall glossy gray with red stripes on the tail. All other markings were black, except for the colorful national insignia and the ***RESCUE*** *markings on the nose. At right is the left side of 160694 as it appeared on May 19, 1978. It has the same scheme and markings as the aircraft shown at left, however, the radome is of a different color.* *(Flightleader)*

Left: F-14A, 160914, in low visibility markings, was photographed in April 1980. This Tomcat carried dark gray stripes and the ***NJ*** *tail code on the tail. Notice the modex of* ***663*** *on the nose and top of the tail.*
(Picciani Aircraft Slides)

F-14A, 161167, of VF-124, was photographed at Dobbins AFB, on January 24, 1982. The stripes on the tail had been changed to a straight style instead of being curved as seen on the previous page. All markings were gray, except for the large, colorful national insignia on the front of the fuselage. (Flightleader)

At left is the left side of F-14A, 160652, that was seen at Offutt AFB on September 19, 1981. The aircraft was overall glossy gray with dark gray markings. At right is a close-up of the tail of the aircraft which gives a better view of the squadron markings. (Both Cockle)

More low visibility markings are seen in this view of 160673 that was taken on May 5, 1984. The squadron markings on the tail had been reduced to two small stripes through the ***NJ*** *tail code. All other markings were black, except for the* ***435*** *modex on the nose and the small national insignia, which were gray. (McGarry via Cockle)*

A close-up of 161606 illustrates the new reduced size of the squadron markings that were applied to the tail. (Grove)

VF-124 operated F-14A, 159827, that had previously been assigned to VX-4. The aircraft was seen on February 19, 1978, in a splinter gray paint scheme. All markings were dark gray, except for the bureau number on the rear of the fuselage, which was light gray. *(Logan)*

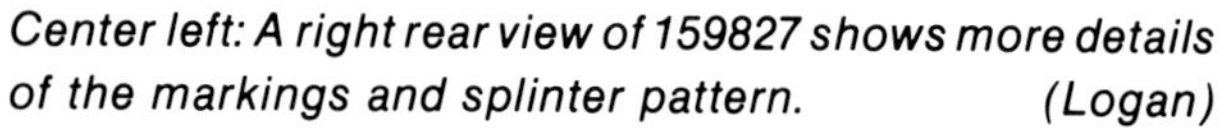

Center left: A right rear view of 159827 shows more details of the markings and splinter pattern. *(Logan)*

Center right: A left side view illustrates the pattern that was painted on that side of the aircraft. *(Logan)*

Right: This close-up of the tail shows the older style of squadron markings. *(Logan)*

Above: VF-124 painted F-14A, 162588, in special markings and used it as a show aircraft. It is painted in the overall gray scheme, and has a special Tomcat insignia on the tail. The ***450*** *modex and* ***50*** *on the tail are in yellow, and are shadowed in black. The black from the radome extends onto and behind the canopy rails. A small, colorful national insignia is located low on the fuselage side. The outstanding unit ribbon is below the canopy, and the Safety* ***S*** *is on the nose below the windscreen. The* ***NJ*** *tail code is located on the inside of the rudder in black. This photograph is dated April 26, 1987, and was taken at MCAS El Centro, California.* *(Nuanez)*

Right: This close-up shows the tail of the aircraft shown above. The colors are visible in the top photo, but this one shows the details better. Note the ***FIGHTERTOWN USA*** *under the emblem. It is the nickname for NAS Miramar.* *(Nuanez)*

These two photographs show one example of the temporary camouflage schemes that were applied to several of VF-124's Tomcats that participated in the Reconnaissance Air Meet at Bergstrom AFB, Texas, in November 1986. The aircraft is painted in the tactical blue/gray scheme, with a water soluble camouflage of dark green and brown applied over the base scheme. It is evident in the photo at right that the scheme was intended to be temporary, and was applied hastily. It would appear that it was applied with a broom or a mop. *(Both Cockle)*

VF-154 BLACK KNIGHTS

*F-14A, 161624, of VF-154, taxies out for a mission in September 1984. The aircraft was an overall glossy gray with black and red stripes on the tail. The **NK** tail code was applied to the inside of the vertical tail surfaces. The ventral fin had a black and red stripe outlining it. The **107** modex on the nose was black, shadowed with red. All other markings were black. Notice the stylized black marking that starts on the radome and continues behind the canopy to a point on the top of the fuselage.* *(Grove)*

Fighter Squadron One Five Pour was commissioned in 1943, and nicknamed the "Black Knights." During World War II, they participated in the battles of Palau, Sipan, Wake Island, and the Philippine Sea. In Korea, while flying the F9F Panther, they expended over 1,500,000 rounds of ammunition and dropped over 470 tons of bombs on enemy positions. On June 15, 1953, the squadron flew a record forty-eight sorties, the most ever launched in a single day by a Navy fighter squadron.

VF-154 launched its first strike in the Vietnam War on February 7, 1965, flying the F-8 Crusader. They returned to NAS Miramar in November of that year to begin transitioning to the F-4B Phantom. They made combat deployments in every year from 1966 to 1970. When the squadron returned from its last combat tour in 1970, it was awarded the Clifton Award as the best fighter squadron in the Navy. In June 1971, they received the F-4J Phantom, and embarked on their seventh and last combat tour, participating in the last Naval strikes of that war. In May 1974, they began their first peace-time cruise in ten years.

VF-154 transitioned to the F-4S in late 1979, but in January 1981, the F-4N replaced the F-4S, and the squadron embarked aboard the **USS Coral Sea, CV-43,** in August. They flew the F-4N for about two and one-half years, then began the transition to the F-14 Tomcat in October 1983. After becoming operational in the Tomcat, they have made cruises as part of CVW-14 aboard the **USS Constellation, CV-64.**

VF-154 also used the tactical blue/gray scheme, as seen in this photograph that was taken in September 1984. All markings were a contrasting gray on the tactical blue/gray paint scheme. *(Grove)*

F-14A, 161624, is illustrated as it appeared at NAS Miramar in February 1986. The markings were the same as seen at the top of the page, except for the stylized black on the nose of the aircraft, which was not present when this photograph was taken. *(M. Geer)*

VF-191 SATAN'S KITTENS

F-14, 161298, was the first Tomcat to be painted in VF-191's new markings. These markings may be tentative, and, at press time for this book, this was the only aircraft painted in the squadron's markings. The aircraft is painted in the overall gray scheme. *(Huston)*

VF-191 was originally formed as VF-19 in July 1943. Flying the famous F6F Hellcat, they began their only cruise of World War II aboard the **USS Lexington, CV-16,** in July 1944. Their success in escorting dive and torpedo bombers is significant in that not one of the aircraft they escorted was ever shot down. After a four month cruise as part of Air Group 19, their tally was 155 enemy aircraft shot down plus five combat ships and twenty-five cargo ships sunk. After World War II they transitioned to the F8F Bearcat, and were redesignated VF-19A, then VF-191.

In 1950, before deployment to Korea, the Blue Angels joined VF-191, thus marking the only time the Blues were to see combat in their history. Following the Korean conflict, the unit flew the FJ-3 Fury, F11F Tiger, and F-8 Crusader. They flew the Crusader during several deployments to Yankee Station off Vietnam, before transitioning to the F-4 Phantom in 1976. They made one cruise with the Phantom before being disestablished on March 1, 1978.

VF-191 was recommissioned on December 4, 1986, and began training with VF-124 to become one of two new F-14 squadrons. Along with the other new squadron, VF-194, they will be part of CVW-10 assigned to the **USS Independence.**

*At left is a right side view of the same aircraft as seen above. Note that the modex is **101,** indicating that it is the squadron commander's aircraft. At right is a close-up of the tail markings. Note the carrier name **USS INDEPENDENCE** on the tail. The large diamond on the rudder is red, and the red stripe at the top of the tail has three white diamonds within. The **NM** is offset, and is in black.* *(Both M. Geer)*

VF-194 HELLFIRES

This photograph, taken on April 29, 1987, shows the CAG aircraft for VF-194. It is painted in the overall gray scheme, and has a red and black modex. The lightning bolt is in red, and a black, offset ***NM*** *tail code is on the rudder.* ***VF-194*** *also appears on the tail, but there is no carrier name.* *(Huston)*

VF-194 had its beginnings as VF-20 in 1942. After seeing combat flying the F6F Hellcat in World War II, they were redesignated VF-9A in 1947, and transitioned to the F8F Bearcat. In 1949 their designation was changed again, this time to VF-91. They saw combat during the Korean Conflict, flying 1938 missions in F9F-2 Panthers from the **USS Philippine Sea, CV-47.**

Following Korea, they flew the F9F-6 Cougar, FJ-3 Fury, and then the F8U Crusader from several carriers including the **USS Hornet, CV-12, USS Kearsarge, CV-33, USS Ticonderoga, CVA-14,** and **USS Ranger, CVA-61.** On August 1, 1963, the squadron was redesignated VF-194, the designation they retain today.

When the war in Vietnam began, the "Red Lightnings" were deployed aboard the **USS Bon Homme Richard, CVA-31** with CVW-19. Later Vietnam deployments were made aboard **USS Ticonderoga** and **USS Oriskany, CVA-34.** The squadron had been the first west coast squadron to qualify in the Crusader and the first Navy squadron to deploy with that aircraft. In March 1976, they were the last operational fighter squadron to fly the F-8, and in that month, they began transition to the F-4J Phantom. Their first cruise with the F-4J was aboard the **USS Coral Sea, CVA-43,** with CVW-15 in February 1977. After that cruise, the squadron was disestablished on March 1, 1978.

On December 1, 1986, VF-194 was recommissioned at NAS Miramar as one of two new F-14 squadrons in CVW-10. At press time for this book, they were still in their initial training period, and had three Tomcats in squadron markings. Once they become fully operational, they will make their first deployment aboard the **USS Independence, CV-62,** when that carrier returns to duty following its Service Life Extension Program (SLEP) yard period.

Earlier markings used by VF-194 are shown in this in-flight shot. This actually is a VF-124 aircraft, and carries the ***NJ*** *tail code. A red lightning bolt is painted through the tail code, and* ***VF-194*** *is painted at the base of the tail. The photograph dates from December 1986.* *(LTJG Jalette)*

The tactical scheme is used on this VF-194 Tomcat. The markings are the same as they were in color, but are in dark gray instead of red. The lightning bolt is outlined in a very light gray. *(Huston)*

VF-211 FIGHTING CHECKMATES

*F-14A, 159626, from VF-211, sits on the ramp at NAS Miramar in full color squadron markings. The aircraft was in the gray over white scheme with a red checkerboard on the rudder. Red, white, and blue horizontal stripes were painted on the top and bottom of the tail. The ventral fin was red, with a white and blue outline, and **VF-211** in white. Notice the red stripe that starts at the nose and outlines the black of the anti-glare panel. The red stripe tapers off to a point behind the cockpit. The carrier name, **USS CONSTELLATION,** is painted in black on the bottom of the engine intake.*
(Bergagnini)

The history of VF-211 dates back to May 1945 when it was formed as VB-74. Since then the squadron has had several designations that include VA-1B, VA-24, VF-24, and the present VF-211, which dates from March 1959. The first fighter flown by the squadron was the F4U-4 Corsair in February 1949, which subsequently gave way to the F9F Cougar, FJ-4 Fury, and the F3H Demon. In early 1959, the F11F Tiger replaced the Demon, and the designation was changed to the present VF-211. Concurrently, the then VF-211, based at NAS Alameda, became the present VF-24. In December 1959, the squadron replaced the Tigers with the F-8 Crusader.

The combat history of the "Fighting Checkmates" includes both Korea and Vietnam. During the Korean Conflict, they flew the Corsair from the **USS Boxer, CV-21.** Later they flew the Corsair, then the F9F-6 Cougar from the **USS Yorktown, CV-10.** In Vietnam, VF-211 destroyed two MiG-17s in aerial combat during a 1965-66 cruise. This made them the first Crusader squadron to down an enemy aircraft. Later they would down four more MiGs, finishing the war with six kills and five probables.

In May 1973, VF-211 made its first peacetime cruise aboard the **USS Hancock, CVA-19.** During that cruise, the October Middle East War and a world-wide military alert gave cause for the carrier and VF-211 to sail to the Arabian Sea, becoming the first carrier to operate there in nine years.

The F-8J was the last version of the Crusader that was flown by the "Fighting Checkmates," and they were traded in for the F-14 Tomcat in October 1975. The squadron "stood up" as a Tomcat unit in December 1975, and was assigned to CVW-9. The first cruise with the F-14 was not until April 1977, and it was made aboard the **USS Constellation, CV-64.** Subsequent cruises were made, with a change in carrier to the **USS Ranger, CV-61.** Today VF-211 remains a part of CVW-9, which is now assigned to the **USS Kitty Hawk.**

This close-up of the nose of 159626 gives an excellent view of the red stripe that runs from the nose to behind the cockpit.
(Kinzey)

*Compare the tail of 159626 with that at the top of this page. The modex **05** had been moved to the inside of the tail in this photo.*
(Kinzey)

*VF-211's CAG aircraft is shown here as it appeared in May 1976. The aircraft was in the gray over white scheme, and carried full color squadron markings. The **100** modex on the nose and the **00** on the tail were the only CAG markings that were carried at that time.* *(Kasulka via Geer)*

Center: Taken in June 1976, this photograph shows the additional CAG markings that were applied to 159636. ***COMCARAIRWING NINE*** *has been added to the area behind the glove vane. Multi-colored stars have been added to the tail around the* ***NG*** *tail code.* *(Flightleader)*

Left: F-14A, 160914, was the new CAG aircraft for VF-211 in June 1981. The aircraft was overall glossy gray, and had full squadron markings. The ***NG*** *tail code was much smaller and was horizontal on the tail, instead of being staggered.* *(Grove)*

VF-211's low visibility markings had made their appearance by the time the photograph at left was taken in September 1978. This aircraft was overall glossy gray, and the squadron markings were the same as seen on the previous page, but the color had been replaced with dark gray. At right is a left side view of 159620 in the same type of markings, as it appeared on October 26, 1980. The red stripe was still carried on the aircraft, extending from the nose to behind the cockpit. All other squadron markings were in dark gray. *(Left Flightleader, right Rogers)*

Right: This close-up illustrates the two styles of markings that were carried by VF-211 in September 1978. The closest aircraft, 159462, carried the squadron markings in full color, while the next aircraft, 159632, had the squadron markings in dark gray. *(Flightleader)*

F-14A, 159467, was photographed in September 1978, in subdued markings. There was no modex carried on the tail of this aircraft at that time. *(Flightleader)*

This F-14A did not have the checkerboard and stripes applied to the inside of the tail as the aircraft in the photograph to the left did. *(Flightleader)*

Seen at Offutt AFB, on May 18, 1984, 161165 was in subdued markings to include the small, gray national insignia on the nose. No carrier name was carried at that time. *(Cockle)*

This close-up of 160930 shows the subdued squadron markings that were applied to the tail. *(Cockle)*

F-14A, 159602, that was photographed at NAS Miramar on May 5, 1984, shows the subdued markings that were carried by another one of VF-211's Tomcats. *(Cockle)*

Another change to low visibility markings is seen in this view of 161168 that was taken at NAS Miramar in October 1986. The stripes have been removed from the front of the aircraft and from the tail. The only squadron markings that remain are the checkerboards on the rudder. The aircraft still carries the large, colorful national insignia. *(M. Geer)*

F-14A, 159637, is shown as it appeared on October 18, 1986. The aircraft was in the tactical blue/gray scheme with darker gray markings. *(Binford)*

VF-211's CAG aircraft was photographed as it taxied out for a mission in July 1986. The aircraft was painted in a brown and dark green camouflage that had been applied over its tactical blue/gray scheme. *(Grove)*

*At left is the right side of **106** that also carried the water soluble camouflage paint. The colors were dark green and brown. At right is the left side of the same aircraft. Part of the checkerboard on the rudder is visible above the camouflage pattern.* *(Both Grove)*

Another one of VF-211's F-14As, that was painted in a temporary camouflage scheme, was photographed as it taxied out for a mission. The only color visible on this side of the aircraft is a dark brown. *(Grove)*

*The other side of **107** is seen in this photograph. Both the dark green and brown were used on this side of the aircraft.* *(Grove)*

VF-213 BLACK LIONS

One of the first aircraft delivered to VF-213 was 159859, the squadron commander's aircraft, seen here at NAS Norfolk, Virginia, on September 25, 1978. The aircraft was in the gray over white scheme, with the squadron colors applied to the rudders and top of the tail. Notice the white ventral fin which was outlined in blue. ***VF-213*** *was painted on the fin in black.* *(Flightleader)*

VF-213 was commissioned at NAS Moffet Field on June 22, 1955. It initially was equipped with the F2H-3 Banshee. They were assigned to Air Wing Twelve, and departed for the Far East on board the **USS Bon Homme Richard, CVA-31,** in early August 1956. Shortly after the cruise, the squadron transitioned to the F4D Skyray. Two WESTPAC cruises were made aboard **USS Lexington, CVA-16,** with the Skyray. Then in March 1960, the squadron exchanged the F4D for the F3H Demon. They deployed aboard the **USS Hancock, CVA-19,** in November 1960. After the cruise, they moved to NAS Miramar in 1961.

The transition to the F-4 Phantom was a bit unusual in that VF-213 received the unique F-4G Navy version of the Phantom (not to be confused with the later USAF F-4G Wild Weasel version). This F-4G had data link equipment that was compatible with shipboard and airborne tactical data systems. The first deployment with the Phantoms was in November 1965 aboard the **USS Kitty Hawk, CV-63.** This was a combat tour in SEA. During this cruise the squadron also compiled another "first" in the evaluation of the Approach Power Compensator. After returning on June 19, 1966, the squadron replaced the F-4G with F-4B Phantoms.

Their second combat tour began in November 1966 aboard the **Kitty Hawk.** During this tour they conducted the first U.S. strike against the MiG bases at Kep. Their third combat cruise began in November 1967, during which they set a record for consecutive days on Yankee Station. The squadron flew an unprecedented 1633 hours in 917 combat sorties. A fourth tour followed in 1969. During this tour VF-213 flew 1897 sorties consisting of 3741 flight hours.

After returning to NAS Miramar in September 1969, the squadron received the F-4J version of the Phantom. In November 1970, the "Black Lions" again deployed to SEA aboard the **Kitty Hawk.** A sixth tour followed in February 1972. In this last combat tour, the squadron spent a record 188 days on the line, with some aircrews accumulating over 300 missions. In all, the squadron flew over 2100 missions and dropped over 1200 tons of ordnance.

In July 1974, the "Black Lions" made a two month deployment aboard the **USS America, CV-66,** but they returned to the **Kitty Hawk** for their next WESTPAC cruise that lasted from May to December 1975. At the conclusion of this deployment, and upon their return to NAS Miramar, VF-213 began its transition to the F-14 Tomcat. They became operational in their new fighters in September 1976.

In October 1977, the first cruise with the Tomcats began, and again the carrier was the **Kitty Hawk.** They remained assigned to CVW-11. But their next deployment was something different, having moved with CVW-11 to the **USS America** for a tour in the Mediterranean. This cruise began in March 1979, and was followed by a second tour to the MED. It was not until late in 1981 that the "Black Lions" returned to the west coast. When they returned to sea in the Pacific, it was aboard the **USS Enterprise, CV(N)-65,** in September 1982. It is with CVW-11 and the **Enterprise** that VF-213 remains assigned today.

This close-up of the tail of 159859 shows details of the squadron markings. The carrier name, ***USS KITTY HAWK,*** *was painted on the bottom of the tail in black.* *(Bergagnini)*

F-14A, 160920, which was painted in the overall glossy gray scheme with colorful squadron markings, was seen at Andrews AFB, on March 15, 1980. There was no tail code or carrier name on the aircraft at that time. *(Miller)*

VF-213's CAG aircraft was seen at Offutt AFB on February 2, 1980. The aircraft was overall glossy gray with colorful squadron markings. The carrier name, ***USS AMERICA,*** *was in black under the lion on the tail.* *(Cockle)*

VF-213's F-14As that were photographed on the ramp at NAS Miramar on May 5, 1984, provide an interesting comparison of markings. The first and third F-14 still carry the squadron markings in full color, while the others were dark gray. Notice the difference in the lion's head between the first and second F-14. The head on the colorful markings was turned to the side, while the gray lion's head was looking straight ahead. *(McGarry via Cockle)*

F-14A, 160910, was photographed in the tactical blue/gray paint scheme. All markings were dark gray except for the black **212** modex on the nose. The carrier name, **USS ENTERPRISE,** was painted on the outside of the tail. The **NH** tail code was applied to the inside of the tail. (Grove)

At left is a close-up of the tail of 159860 as it appeared at Dobbins AFB on September 20, 1986. The **NH** tail code was on the outside of the tail in smaller letters than had been used on the aircraft in the photograph at the top of this page. The gray lion was different from the one that was used with the colorful markings. The stars on the rudder were a dark gray. The **NAVY, VF-213,** and the bureau number were black. At right is an overall view of **204** that shows the tactical blue/gray scheme to good effect. (Both Cockle)

F-14A, 160920, was photographed at NAS Miramar on March 26, 1987, in the tactical blue/gray paint scheme. (Kinzey)

A close-up of the tail of 160920 shows the subdued markings that were painted there. (Kinzey)

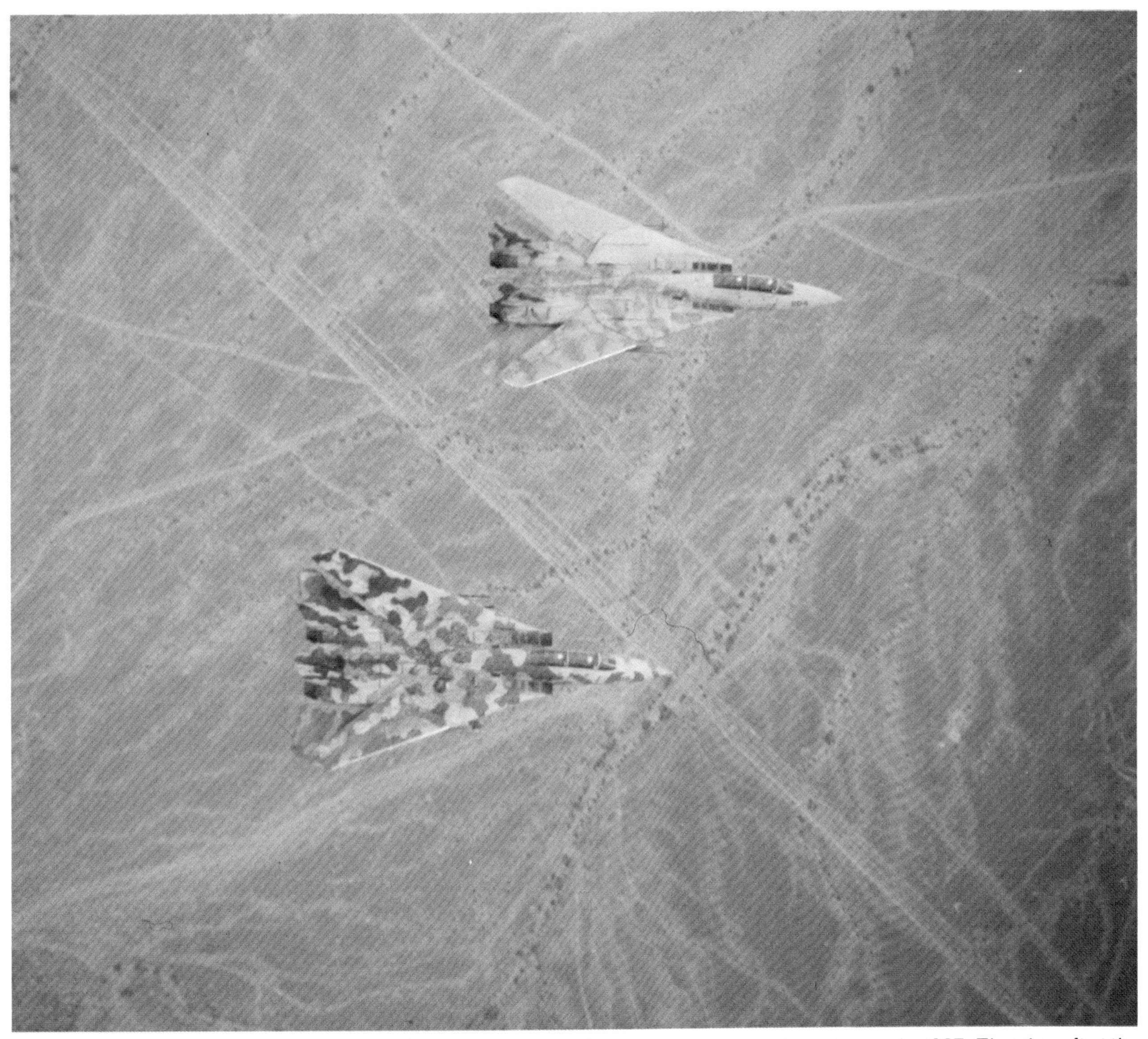

Two of VF-213's F-14As are seen here in the temporary paint schemes that were tested in early 1987. The aircraft at the top of the picture carried a brown scheme in a diagonal pattern across half of the tactical blue/gray scheme. The aircraft at the bottom carried a gray and blue/gray pattern over the tactical scheme. (LT Pollard)

Another F-14A of VF-213 is illustrated as it appeared during the camouflage test in early 1987. The ***212*** *modex is visible on the nose, and portions of the squadron's markings are visible on the tail under the dark gray temporary camouflage pattern. (LT Pollard)*

Tomcat ***210*** *was photographed as it taxied out for a mission at NAS Miramar. The aircraft carried a dark green and brown camouflage pattern over the tactical blue/gray scheme. (LT Alcala)*

*Above: This beautiful in-flight shot of **210** shows the camouflage pattern that was applied to the left side of the aircraft. The **NH** tail code and the lion are just visible under the brown paint on the tail.* *(LT Pollard)*

Left: VF-213's CAG aircraft was photographed over the California desert, with a dark gray pattern over the tactical blue/gray scheme. *(LT Pollard)*

At left is an in-flight photo of the underside of the CAG aircraft that shows the camouflage pattern that was applied there. At right is a rear view of the same aircraft on the wash rack at NAS Miramar. The camouflage had started to wear off when this photograph was taken on March 26, 1987. While the paint was a water soluble type, it was surprisingly difficult to remove. *(Left LT Pollard, right Kinzey)*

VF-301 DEVIL'S DISCIPLES

F-14A, 158990, of VF-301, which is part of Navy Reserve Wing 30, was seen at NAS Miramar in September 1966. The aircraft was overall gray with black markings. The arrow on the tail was red, outlined in black. *(M. Geer)*

It was January 3, 1944, when VF-301 was first commissioned, being initially equipped with the F4U-1 Corsair. Their World War II history was quite short, and included a deployment aboard the **USS Steamer Bay** and a basing at Luganville Airfield, Esperitu. The unit was disestablished August 1, 1944.

It was not until October 1, 1970, that the squadron was reestablished at NAS Miramar as a reserve fighter squadron. At that time they were equipped with another Vought product, the F-8 Crusader. They were the first of two reserve squadrons at Miramar, and were assigned to CVWR-30. After almost four years, they turned in their F-8s for the F-4B Phantom. These were replaced by the F-4N in February 1975, followed by the F-4S in November 1980.

Although the first F-14 was received by VF-301 in October 1984, the transition was somewhat lengthy, and the F-4S remained in service with the unit well into 1985. Today, the "Devil's Disciples" are fully equipped with the Tomcat, and although not a sea-going unit, they maintain proficiency in their aircraft to include carrier qualifications. Their training periods are designed to keep their personnel and aircraft ready to answer any call in the event of national need.

Center: F-14A, 159442, is painted in the tactical blue/gray paint scheme. The markings were dark gray to include the outline of the arrow on the tail. The ***113*** *modex on the nose was black.* *(M. Geer)*

Right: F-14A, 159085, is illustrated in the tactical blue/gray paint scheme on March 8, 1986. All markings were a contrasting gray over the tactical blue/gray scheme. *(Slowiak)*

VF-302 STALLIONS

*VF-302's CAG aircraft, 158989, was photographed in colorful markings in June 1986. The aircraft was in an overall glossy gray scheme, and had a large, gray national insignia on it. The squadron markings were on the tail in yellow, outlined in black. The horse-head figure on the rudder was black, as was the **ND** tail code. The ventral fin had a yellow and black stripe outlining it, with **VF-302** in black. The canopy had a glossy black area painted around it which was outlined in yellow. The crew names on the canopy rail were yellow.* *(Grove)*

The second reserve unit at NAS Miramar is VF-302. It was commissioned on May 21, 1971, and like VF-301, is assigned to CVWR-30. It operated the F-8K until November 1973, at which time it transitioned to the F-4B. The F-4N replaced the F-4B in January 1975. The unit was presented with the F. Trubee Davidson Award as the most outstanding "tailhook squadron" in the Naval Air Reserve in fiscal year 1977. The following year they took the Top Squadron trophy for their victory in the first annual Reserve Fighter Derby. Next came the Battle "E" signifying the "Stallions" as the best fighter squadron in the Naval Air Reserve Force. In the summer of 1981, the F-4S replaced the F-4N as the mount of VF-302. Several Safety "S" awards were won by the squadron, as was the Golden Tailhook award and the F. Trubee Davidson Award again in 1981. The award was repeated in 1984.

Beginning in February 1985, VF-302 began transitioning to the F-14 Tomcat, and made the shortest transition in the history of the F-14. In January 1986, they deployed aboard the **USS Ranger, CV-61,** for intensive training in carrier day and night operations. Later that year they became the first Reserve TARPS squadron.

Center: Here are two of VF-302's F-14As that took part in the Reconnaissance Air Meet 86 at Bergstrom AFB. The aircraft are seen here doing a quick check prior to take off. The aircraft were tactical blue/gray with dark gray markings. *(Cockle)*

*Right: F-14A, 158984, was photographed as it taxied out for a flight in August 1985. The aircraft was overall tactical blue/gray with darker gray markings. The squadron stripe has been removed from the tail, leaving only the horse head and the **ND** tail code on the rudder.* *(Grove)*

VX-4 THE EVALUATORS

*F-14A, 159424, that was assigned to VX-4, is seen at NAS Pt. Mugu, California, on May 21, 1977. The aircraft was in the gray over white scheme, and carried a dark blue band, outlined in red, on the front of the fuselage and on top of the tail. The blue band had white stars painted on it. The **XF** tail code was black, shadowed in white, as was the **34** modex on the nose. Visible under the fuselage are additional ordnance racks that had been added for testing the delivery of air-to-ground ordnance. This modification was not adopted for operational units. (Flightleader)*

VX-4 is not an F-14 squadron in the sense that the other squadrons covered in this book are, but they do operate several F-14s in performing their mission of evaluating and testing air-launched guided missiles, aircraft, and their associated systems. They were formed on September 15, 1952, at Point Mugu, California, where they remain today. The "Evaluators" are assigned to the Navy Test and Evaluation Force. In addition to their evaluation mission, they are also the source of tactical and operational information for the fleet squadrons. They also prepare the tactical manuals for all Navy fighter aircraft.

The first F-14 came to VX-4 in 1972, and the Tomcat has been among the aircraft in the unit's inventory ever since. They have worn some colorful markings, and have often been adorned with the Playboy bunny logo.

F-14A, 158621, is shown as it appeared in October 1973. This aircraft carried the squadron stripes on the tail, but not on the fuselage. (Cockle)

This rear view of 158618 provides a good look at the blue stripes that were applied to the upper portion of the horizontal tail surfaces. The stripes were also a dark blue, outlined with red with white stars. (Cockle)

The Playboy bunny has adorned many of VX-4's aircraft. On this page is a look at several of their F-14s to which this famous logo has been applied. At left is F-14A, 159830, as it appeared at NAF Andrews in August 1983. Notice that no tail code was carried, and that both the old, large, colorful national insignia and the small, gray insignia were present at the same time. A small Playboy bunny was carried on the bottom of the rudder in black. At right is 159828 in the tactical blue/gray paint scheme with contrasting gray markings. The Playboy bunny was carried on the bottom of the rudder in gray. *(Left Brown, right GB Aircraft Slides)*

*Another Tomcat in VX-4 markings, 161444, was seen at NAF Andrews on October 19, 1985. The aircraft was in the overall glossy gray scheme with the squadron stripes on the tail in full color. The **XF** tail code was black, as were the other markings on the aircraft, to include the Playboy bunny on the bottom of the rudder. Notice the AIM-54C training missile that was carried on one of the forward fuselage stations.* *(McGarry via Cockle)*

*F-14A, 159853, was also seen at NAF Andrews on October 19, 1985. This aircraft was in the tactical blue/gray scheme, with contrasting gray markings. The Playboy bunny on the tail was much larger than seen before, and the **XF** tail code had been moved to the squadron stripe at the top of the tail. This aircraft also carried AIM-54C training missiles.* *(McGarry via Cockle)*

*VX-4 had another variation of their markings which was seen on F-14A, 159424. It was photographed at NAS Pensacola, Florida, on May 9, 1986. The aircraft was in the tactical blue/gray scheme, and the squadron stripe on the top of the tail was in full color. The **XF** tail code was superimposed on the squadron stripe in white. The large Playboy bunny on the tail was black, as were the other markings of the aircraft. The small national insignia on the front of the fuselage was gray. The most interesting scheme to carry the Playboy bunny can be seen on page 3.* *(Flightleader)*

This nice landing shot of F-14A, 159830, was taken in September 1978. The aircraft was in a splinter camouflage pattern, consisting of light and dark grays. All markings on this Tomcat were black. (Picciani Aircraft Slides)

*At left is the right side of 159830 as photographed at Nellis AFB, Nevada, in November 1977. The false canopy that was painted on the bottom of the aircraft is partially visible on the nose gear door. At right is F-14A, 159825, which was also in a splinter camouflage pattern. The **VX-4** on the ventral fin and the **40** on the top of the tail were light gray. The **40** modex on the nose was black, as were the crew names on the canopy rail. (Left Logan, right Flightleader)*

At left is the right side of 159827, which was another of VX-4's F-14As that carried the gray splinter camouflage paint scheme. At right is 159829 in the splinter camouflage pattern. There was no tail code carried on either of these aircraft during this time period. (Left Flightleader, right Roth)

F-14A, 159831, is seen in a reddish-brown and olive green camouflage pattern that has been applied over the normal paint scheme. The only marking visible is the ***45*** *modex on the nose in black.* *(Grove)*

Left: F-14A, 159830, was photographed in May 1981. The aircraft was in the overall glossy gray scheme with black markings. The large, colorful national insignia was still being used on this aircraft. *(Flightleader)*

At left is the right side of F-14A, 159424, while it was assigned to VX-4. The aircraft was seen at Dobbins AFB on March 21, 1981, in the overall gray scheme with black markings. The ***XF*** *tail code was black, shadowed with white. At right is the other side of the aircraft as photographed in November 1980. Notice the black* ***F-18 HORNET*** *marking that was carried on the tail.* *(Left Flightleader, right GB Aircraft Slides)*

PMTC/NMC

*Above: This photograph was taken in July 1972, and shows F-14A, 157983, that was used in the Phoenix missile test at NAS Pt. Mugu. The aircraft was in the gray over white scheme with black markings. Notice the last four numbers of the bureau number that were carried on the tail below the Phoenix Missile Test badge. The numbers **7983** were painted in large, black numbers. (Geer)*

*Right: The same aircraft as seen above was photographed at NAS Pt. Mugu in November 1973, in Navy Missile Center markings. The markings on the tail consisted of the Navy Missile Center badge superimposed over a red and a dark blue chevron. **NMC** was painted in large red letters under the badge on the tail. (Roth)*

*The Navy Missile Center was later replaced by the Pacific Missile Test Center. This F-14A, 157988, was photographed on November 1, 1976. The letters **PMTC** were painted in black on the tail above the center's badge. The center's badge was a dark blue, gold, and white triangle. It was outlined in white, and was centered on a light blue horizontal stripe. The **209** modex on the nose was black, shadowed with white. (Knowles)*

Taken in May 1977, this photograph shows F-14A, 158615, in PMTC markings. These are the standard markings used by the Center at that time. (Flightleader)

This Tomcat was painted in the gray splinter camouflage that VX-4 had tested at one time. The markings were gray, except for the PMTC badge and the blue tail stripe. The **201** modex on the nose was black, shadowed with white. (Flightleader)

The overall glossy gray scheme is illustrated on 158625 in this photograph dated October 23, 1983. The Center's badge was carried on the blue stripe, and was in full color. The **26** modex was in black at the top of the rudder, and **PMTC** was in black above the badge. The **226** modex was on the nose in black, shadowed in white. (Kaston)

This left rear view of 158623 clearly shows the markings that were applied to the tail of the aircraft. Notice that the **24** modex was painted on top of the tail instead of the top of the rudder. (Grove)

THE MOVIE "TOP GUN"

*Above: One of the F-14As that was used in the movie "TOP GUN" is seen in this photograph. The aircraft was in the tactical blue/gray scheme, and carried a fictitious squadron badge for VF-1 in full color on the tail. All other markings on the aircraft were gray, except for **NAVY** on the fuselage and the **114** modex on the nose, which were black.* *(LT Baranek)*

*Right: The tail of another "TOP GUN" F-14A is just visible on VF-114's flightline. The badge is badly weathered, but still present. **VF-1** was painted in black on the ventral fin.* *(Kaston)*

*At left is a close-up of the tail of F-14A, 160681, that was also used in the movie. The aircraft was assigned to VF-111, and carried a squadron badge for VFA-213. This was a fictitious squadron used in the movie. The badge would appear to be based on the markings of VA-25. It consisted of a black fist holding a red lightning bolt. The circle was yellow, outlined in black. There were three black stars in the yellow circle. A black **VFA-213** was painted on the yellow ribbon, which was outlined in black. At right is a closer view of this squadron badge.* *(Both Kaston)*

NASA

F-14A, 157991, was photographed at Edwards AFB, California, in November 1982. At that time the aircraft was on loan to NASA. It was in the gray over white scheme, and had red-orange horizontal and vertical tail surfaces, and outer wing panels. Notice the black canard surfaces that had been added to the nose in front of the canopy. ***NASA*** *and* ***991*** *were painted in black on the top of the vertical tail.* *(Flightleader)*

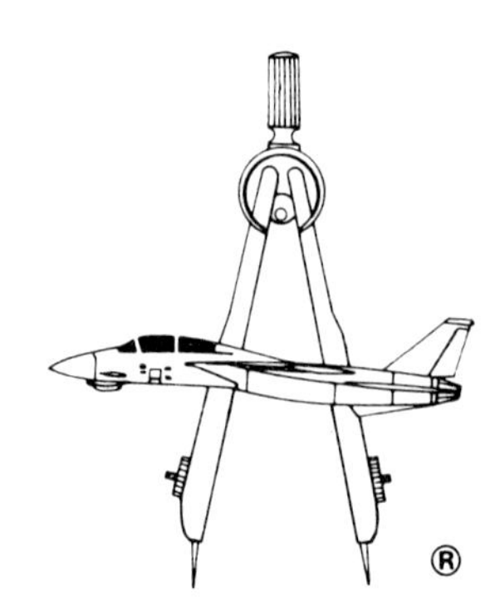